Disaster Zone

The Science Behind

SUPER FLOODS

Louise Spilsbury

CHERITON
CHILDREN'S BOOKS

Please visit our website, www.cheritonchildrensbooks.com to see more of our high-quality books.

First Edition

Published in 2022 by **Cheriton Children's Books**
PO Box 7258, Bridgnorth WV16 9ET, UK

Author: Louise Spilsbury
Designer: Paul Myerscough
Editor: Tim Cooke
Proofreader: Wendy Scavuzzo
Picture Researcher: Rachel Blount
Consultant: David Hawksett, BSc

Picture credits: Cover: Shutterstock/Olha Tytska; Inside: p1: Shutterstock/AJP; pp4-5: Shutterstock/Topten22photo; p5r: Shutterstock/CEW; p6: Shutterstock/Zixia; p7: Shutterstock/Roschetzky Photography; p8: Shutterstock/Lucky Team Studio; pp8-9: Shutterstock/Zack Frank; pp10-11: Wikimedia Commons/U.S. National Archives and Records Administration; p11r: Wikimedia Commons/U.S. National Archives and Records Administration; pp12-13: Shutterstock/Annalisa e marina Durante; p12b: Shutterstock/McAuley Multimedia; p14: Shutterstock/Steve Heap; p15: Shutterstock/Ryan Morgan; p16: Shutterstock/FCG; p17: Wikimedia Commons/George Armstrong/FEMA; pp18-19: Shutterstock/Everett Historical; p19r: Wikimedia Commons; pp20-21: Shutterstock/Mike Ver Sprill; p21t: Shutterstock/DmyTo; p22: Shutterstock/AJP; p23: Shutterstock/Michelmond; p24: Shutterstock/Lienka; p25: Shutterstock/Steven Collins; pp26-27: Shutterstock/Kev Gregory; p27b: Shutterstock/AJP; pp28-29: Wikimedia Commons/NOAA/NASA GOES Project; p29b: Shutterstock/Andrey VP; p30: Shutterstock/Audrius Merfeldas; p31: NASA/JPL/QuikScat Science Team; pp32-33: Wikimedia Commons/NOAA/NASA/UWM-SSEC-CIMSS/ William Straka III; p33r: Shutterstock/MicheleMidnight; pp34-35: Shutterstock/Lexaarts; p35r: Wikimedia Commons/NASA scientist Joey Comiso; p36: Wikimedia Commons/Matt Affolter; p37: Wikimedia Commons/NASA/JPL/Corby Waste; pp38-39: Shutterstock/Media Bakery13; p39r: Shutterstock/Nomad_Soul; pp40-41: Shutterstock/IR Stone; p41r: Shutterstock/Aviahuisman; p42: Shutterstock/ Hank Shiffman; p43: Shutterstock/Herdik Herlambang; p44: U.S. Air Force photo/Staff Sgt. Kelly Goonan; p45: Shutterstock/Maythee Voran; p47: Shutterstock/AJP.

Printed in the United States of America

Contents

Chapter 1

FEARSOME FLOODS

Floods are a common natural disaster. Throughout history, they have caused terrible damage. In the last 100 years alone, they have killed millions.

What Is a Flood?

A flood is when an area of dry land is covered by water. Plants, people, and other living things need water to live. When floods slowly cover fields next to a river every year, this is a good thing. The floods deposit rich soil and water the plants that grow there. However, when a flood happens where people live, it can be very destructive. Some floods are small or form slowly. They take hours or even days to develop, so people have a lot of time to clear their houses. They can move their precious possessions and themselves to safety, out of reach of the water. Sometimes, floods happen so quickly that people have no time to escape. That is when a disaster happens.

When a flood covers land that is normally dry, it causes **chaos**. It can wash away people, vehicles, and trees. It can bury houses under mud. It can ruin **crops** and destroy bridges and roads.

Flash Floods

Floods that form quickly and with little warning are called flash floods. Flash floods can be very dangerous—they can turn a gentle stream into a huge flow of water. A flash flood can be so powerful that it sweeps away everything in its path. People, cars, trees, and even houses can be washed away. Flash floods happen when a lot of heavy rain flows down hillsides or mountainsides into valleys. Sometimes, these floods hit an area far from where it is raining.

Flash floods are dangerous because there is often very little time between the rain falling and floods occurring.

SUPER FLOODS

Once in a while, a flood unlike any other comes along and releases a huge amount of water. That type of flood is called a mega or super flood. Mega floods are rare but, when they do occur, they have so much **energy** that they destroy everything in their path. At the end of the last **ice age**, the planet began to get warmer. That caused **ice sheets** and **glaciers** to melt, resulting in super floods.

DEADLY DISASTERS

Floodwater can kill people and destroy buildings. When a flood hits an area without any warning, people can be swept away by its force. The walls of mud that the water washes off the land can also kill.

The Power of Moving Water

In a flash flood, water rushes through streets and houses very quickly. Moving water is very powerful. As the water moves through a town, it knocks down bridges and damages buildings. It leaves behind thick mud that buries houses. It is hard to get out of the way of water that moves so quickly, and many people die because they make the mistake of trying to drive or walk through floodwater. People who cannot move away quickly enough can drown.

These smashed cars and piled-up **debris** are the result of heavy flash floods in a village in Mallorca, Spain, in 2018.

Where in the World?

In 2013, very heavy rain fell in the mountains of Colorado. The rainfall filled creeks and rivers, including the Colorado River. The river overflowed, causing a terrible flood. Floodwaters poured over roads and bridges, across neighborhoods and farm fields. Boulder, Weld, and Larimer counties had the most damage. The floods killed 9 people and destroyed 1,852 homes. State highways the length of 120 miles (193 km) were destroyed or damaged. The flood caused $4 billion in damage.

The Colorado River flooded due to a heavy rain storm in the mountains.

Dangerous Debris

A big problem with fast-moving water is that it picks up and carries objects along with it. The speed of the water turns those objects into deadly weapons. Floodwaters carry trees, rocks, vehicles, and even parts of roads, bridges, and buildings, which have been broken up by the force of the waters. These dangerous objects can badly hurt or kill people in or at the edge of the water if they hit them. Floodwater of just 24 inches (61 cm) is enough to pick up and carry along a car or a truck.

Deadly Fire

Water can also knock down power lines, and break gas and oil pipes. If a person touches water after a broken power line has fallen into it, they might die. That's because the water is full of electricity from the power lines. When broken power lines touch each other, they send out sparks that can set fire to buildings. Broken gas pipes also cause explosions and fires.

A TRAIL OF DESTRUCTION

When floodwaters finally drain away, people are left with a lot of problems. The damage big floods cause can take months, and sometimes years, to repair and recover from.

In October 2009, floods and mudslides in Sicily, Italy, caused chaos and devastation.

Damage to the Environment

A flood can completely change the land. Floods can rip soil out of the land and leave behind large holes. The soil in floodwaters can make water supplies dirty. Also, it is possible for chemicals and other dangerous substances to get into the water, which can poison it. That can lead to diseases that could kill people. **Dysentery** is a disease people get when they drink or use dirty water. If dead animals or moldy food are in floodwater, bugs will grow. Those bugs are **viruses** and **bacteria** that lead to disease.

The Road to Recovery

It can take years to recover from a flood because people have to rebuild their homes, businesses, and lives. For example, when a flood destroys roads, railroads, and bridges, it can be a long time before people are able to travel anywhere. When power lines are damaged, the electricity goes off and without electricity, computers and other tools do not work. That makes it difficult for people to work. If homes and farms are destroyed by the water, people are left with nowhere to live. It takes a long time to clear mud and all the trash from the streets. Also, it will take a long time to repair and rebuild damaged property. When farm crops are destroyed, there may not be enough food for everyone.

Where in the World?

About 15,000 years ago, tons of water poured across part of eastern Washington State. In just a few days, the landscape changed. The force of the flood ripped soil and rock out of the ground and dug out deep **canyons**. It also loosened a huge ice sheet that moved to block nearby lakes and rivers. They overflowed, creating a huge waterfall that was eight times taller than Niagara Falls. Piles of gravel as high as skyscrapers were picked up and dropped, and rocks were thrown across the ground. Once the ice melted, the water again flowed normally. It left behind a huge dry waterfall.

The strange-looking landscape is known as the Scablands. It was created by the flood that hit Washington State 15,000 years ago.

Q

We have learned about the physical damage that floods cause. What emotional damage do you think people suffer as a result of floods?

Disaster Zone

Mississippi

The Great Mississippi Flood of 1927 is still the biggest flood in American history, and one of the worst natural disasters to hit the United States. More than 23,000 square miles (59,570 sq km) of land was flooded. Thirteen deep cracks split the ground, hundreds of thousands of people lost their homes, and at least 250 were killed.

A River Overflows

By April 1927, it had been raining for several months in Mississippi. The rainfall was 10 times heavier than usual, and the Mississippi River was full. In Vicksburg, the river had spread to 80 miles (129 km) wide. People could see the river was overflowing, but they still felt safe. The US Army had built special walls, called levees, to hold back the river to protect them.

This photo of the Mississippi River was taken after the flood of 1927. The land next to the river is completely flooded.

Lost Levees

On April 16, the first of the levees broke open. This was followed by a second levee five days later. Over the next few weeks, all of the special walls broke open and fell apart, and millions of gallons of water poured out. Soon the streets were flooded. In some places, homes disappeared under 30 feet (9 m) of water. Tens of thousands of people climbed onto their roofs to escape the water. Other people climbed trees to wait for help. The floodwater stayed until August.

WASHED AWAY

The Mississippi flood became known for the poor treatment of African American **plantation** workers. They had to pile up sandbags on the levees to make them stronger. This photograph shows some plantation workers at work. When the walls broke, the water swept away many of the workers. Thousands of other workers were left for days with no food or water, while white people were rescued and taken to safety. The death toll was listed as 250, but it was likely nearer to 1,000 because many African American deaths were never written down.

The Mississippi flood left African American communities devastated.

Chapter 2

HOW SUPER FLOODS WORK

Heavy rainfall causes most flooding. When a lot of rain falls in a very short time, it cannot all drain away. It builds into floodwater. Most floods happen for this reason, no matter how big or small.

Each time there are more rains than a **drainage system** can take, there can be floods.

The Water Cycle

Rain is part of the water cycle. The water cycle is the movement of water between the **atmosphere** and Earth's surface. The sun's heat makes water disappear into the air from Earth's oceans, seas, lakes, and rivers. When that happens, water turns from liquid into **water vapor**. This is called evaporation. As water vapor rises higher in the sky, where it is colder, the water vapor cools. When it cools, it gets smaller. It turns from a gas into a liquid and forms droplets of water. This is called condensation. A lot of water droplets form clouds. As clouds rise higher, they become even colder. The droplets in them grow into heavier raindrops. When these drops become big and heavy, the air can no longer hold them up. Then, they fall to the ground as rain.

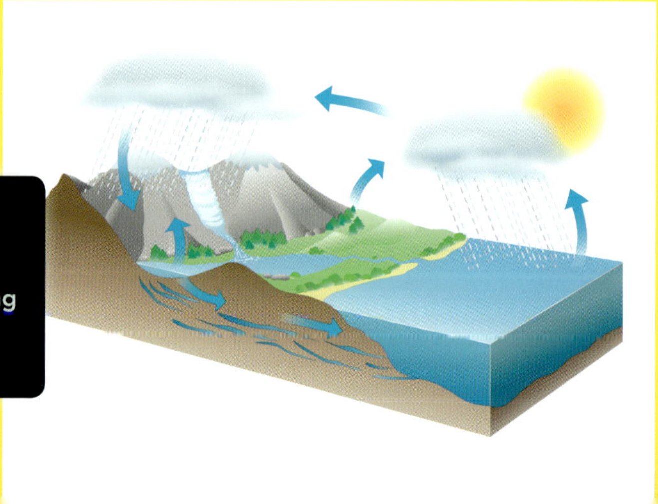

This diagram shows the water cycle. The arrows pointing upward indicate water evaporating, then condensing into droplets in clouds. Arrows pointing downward indicate water droplets growing larger and falling as rain.

Raining Heavily

Heavy rains, or rainstorms, usually happen during hot weather. That is because the warm weather causes more water to evaporate from the land and oceans. The water droplets in the air form larger rain clouds that drop more rain than usual. In a heavy rainstorm, the raindrops are bigger, and there are more of them. That means a lot of water can fall from the sky very fast. Usually, as rain falls, it has time to soak into the soil or flow into drains and rivers that carry the extra water away. When a lot of rain falls, the drains cannot deal with the larger amount of water. The water does not drain as quickly as it should. The drains back up, the water rises, and causes floods.

Where in the World?

In 1861, farmers in California were facing a terrible **drought** because it had not rained for a long time. In December, there was a storm and it started to rain. The farmers thought the drought was over. However, after the storm, the rain did not stop for 43 days, and the heavy rains caused mega floods. Much of California was underwater for almost six months. The floods killed thousands of people and the damage left the state with no money.

Q

What do you think farmers in California today should learn from the mega flood that happened in 1861?

THE RISKS OF RUNOFF

When heavy rain falls on the ground in a short space of time, or water collects in dry places, it cannot drain away or soak into the ground fast enough. It runs off the surface, and this known as runoff.

Snowmelt Floods

When snow melts, it turns to water. This is snowmelt. In many very cold places, a lot of snow falls during the winter. Over the months, the snow builds up. Snow can stay frozen for a long time. If the temperature remains below freezing in the spring, the snow will stay frozen. As it starts to get warmer in the summer, large amounts of ice and snow can quickly melt and turn to water. Places that are normally dry will suddenly have a lot of water flowing through them. This can lead to sudden, heavy snowmelt floods.

Huge amounts of water can flow from snowmelt in mountains, sometimes causing floods.

In 2013, this river overflowed its banks in Alberta, Canada, causing flooding.

BUILDING SLOWLY

It can take a long time for rivers to fill up with water from runoffs. That's because it takes time for the water to find its way to the river. That type of flooding takes longer to happen. However, it can last for days or weeks. It can also spread a long way from the river.

Overflowing Rivers

Snowmelt also flows into rivers. Soon, the rivers' water levels rise and, when they overflow their banks, they flood the nearby land. Rivers also overflow and cause floods when there is a lot of heavy rain in hilly areas. The rain pours down into valleys and land below. When it rains a lot, the soil near a river can become so full and wet that no more rain can soak into it. This is called saturation. When the soil becomes saturated, the rainwater flows straight off the soil and into the river. If there is a lot of this runoff, the river will fill up and overflow its banks.

DAM BUSTERS

Terrible floods can also happen when a dam breaks. A burst dam is very dangerous because suddenly a lot of fast-moving water pours out at once.

The Danger of Dams

A dam is a solid barrier built across a river to stop its flow. The water behind the dam can be stored as a **reservoir**. This water has many different uses, for example, it can water local fields or provide drinking water for local towns. Some dams are used to make electricity for **power plants**. Sometimes, the **pressure** on the barrier can become too great and can make the dam break apart. When that happens, the water pours out into the surrounding area and **downstream**.

When dams overflow after heavy rains, dangerous floods can follow.

In Nevada in 2008, heavy rains broke levees and drainage canals. This caused massive floods that filled local homes.

Broken Levees

Like dams, levees can also break. These are walls built along a river or the sea to stop water from flooding nearby land. If they break, all the water behind the walls is suddenly free. The water flows out so fast, it quickly floods the towns and cities the walls were protecting. On January 5, 2008, a wall broke along a canal by Fernley, Nevada. Six feet (1.8 m) of water soon filled houses, and 3,500 people had to be rescued.

Where in the World?

On May 19, 2020, after a period of heavy rainfall, the Edenville Dam on the Tittabawassee River, Michigan, failed. This led to a flood, which then caused the downstream Sanford Dam to overflow. Some nearby homes were flooded by up to nine feet (2.7 m) of water. A state of emergency was announced and 10,000 people were **evacuated**. Buildings, bridges, and roads were swept away. The failure of the dams was blamed on poor maintenance over many years, followed by heavy rains. The total cost of repair after the flood was more than $240 million.

Disaster Zone

Pennsylvania

It was a cold, rainy day in May 1889, and it had rained heavily all night. The residents of Johnstown, Pennsylvania, suddenly heard a loud noise. The sound grew to a roar like thunder. Some people knew right away what had happened—the South Fork Dam, 14 miles (22.5 km) upstream of the city, had burst.

These houses were ruined by the devastating Johnstown Flood of May 31, 1889.

Disaster Strikes

When the dam burst, it sent a 30-foot- (9 m) high wall of water rushing down a narrow valley. Moving fast at 20 to 40 miles per hour (32 to 64 kph), the water headed for the city. 20 million tons (18 million mt) of water, carrying trees, rocks, barbed wire, and other junk, crashed onto the city's streets. The floodwaters destroyed everything in their path. They tore up steel mills, farm animals, and people. Some homes were ripped apart. Then the flood carried the houses, along with the rest of the debris, hundreds of feet away.

After the Flood

The flood devastated the city. Whole blocks of buildings were ripped apart. Huge piles of mud and rubble filled the streets. The debris reached as high as the third story of buildings. Hundreds of people were buried under the rubble. Four square miles (10 sq km) of the city of Johnstown was completely destroyed. It took five years to clean up. Months and even years later, bodies were still being found. The final death toll of the disaster was 2,209 people.

FROM FLOOD TO FIRE

The flood rushed through Johnstown, carrying away everything, including trees, animals, buildings, freight cars, and even people. It was only stopped by a huge stone bridge at the far side of town. Thousands of tons of debris piled up against the bridge. Oil from machinery and locomotives spilled into the water. The oil suddenly caught fire. Eighty people died at the bridge.

This drawing shows how rescuers fought to free people as the Johnstown flames spread.

Chapter 3

WHERE SUPER FLOODS HAPPEN

A flood can happen just about anywhere it rains. When there is heavy rain, there is always a risk of flooding. Some places flood more often than other places because the water there cannot drain away.

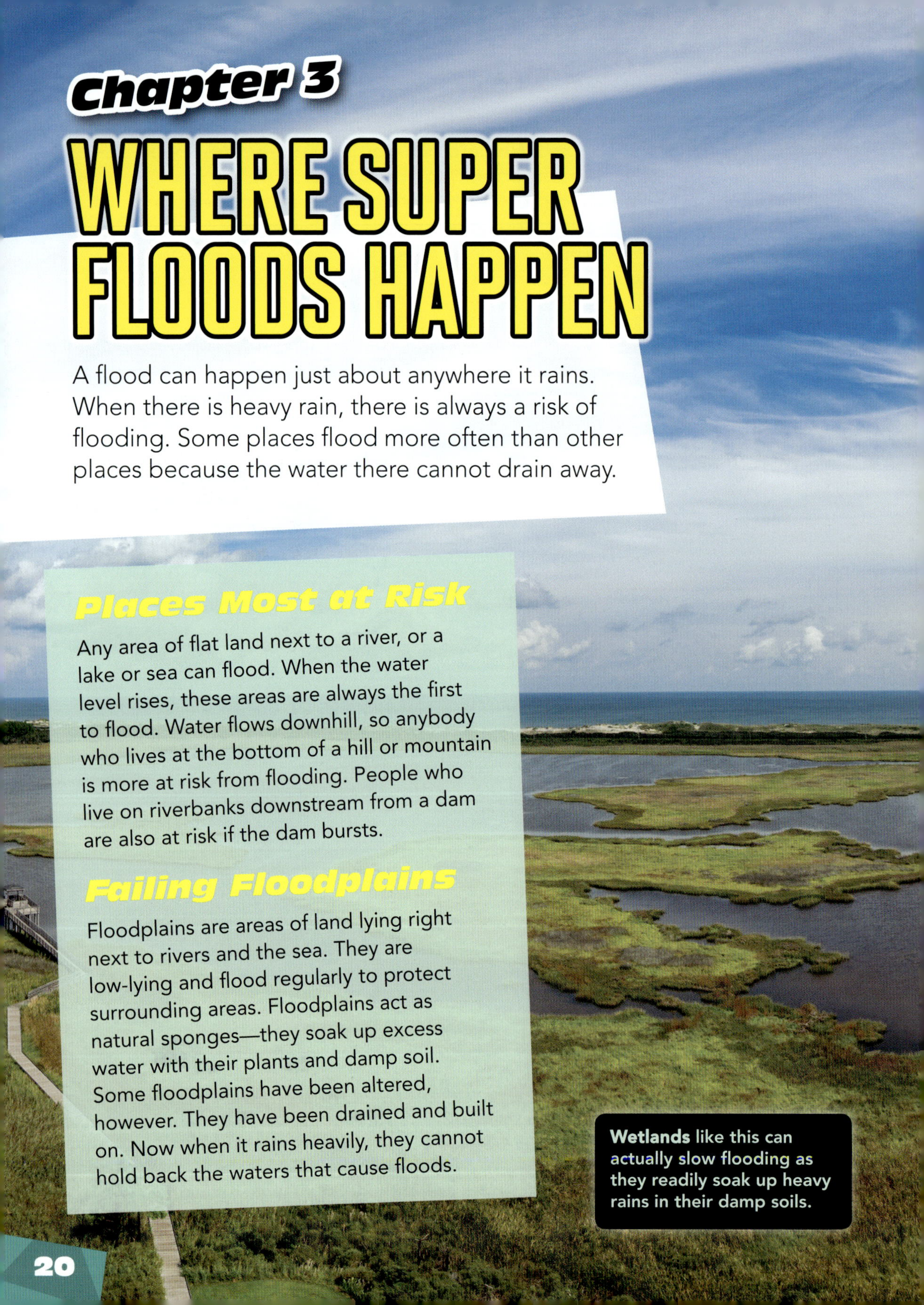

Places Most at Risk

Any area of flat land next to a river, or a lake or sea can flood. When the water level rises, these areas are always the first to flood. Water flows downhill, so anybody who lives at the bottom of a hill or mountain is more at risk from flooding. People who live on riverbanks downstream from a dam are also at risk if the dam bursts.

Failing Floodplains

Floodplains are areas of land lying right next to rivers and the sea. They are low-lying and flood regularly to protect surrounding areas. Floodplains act as natural sponges—they soak up excess water with their plants and damp soil. Some floodplains have been altered, however. They have been drained and built on. Now when it rains heavily, they cannot hold back the waters that cause floods.

Wetlands like this can actually slow flooding as they readily soak up heavy rains in their damp soils.

The Problem of Concrete

Soil is covered up with layers of concrete to build roads and buildings. More concrete means there is less soil to absorb, or soak up, rainwater. Water cannot pass through concrete, but runs off the surface into drains. The more concrete there is, the more water runs off into drains during heavy rains. When drains fill up too quickly, they overflow and the extra water floods the streets.

Water runs off concrete quickly and can cause sudden and deadly flash floods, especially in built-up areas.

Q What measures might we put in place to help deal with the problem of concrete and floods in our towns and cities?

FORESTS AND FLOODING

Deforestation is the cutting down of forests. This increases the risk of flooding, because after trees are cut down, the wind blows away the top layer of soil. This layer's job is to soak up and store rainwater. Without the topsoil, the rainwater runs more quickly off the land. When trees are cut from hills and mountains, rainwater rushes downhill and causes flooding.

FLOOD SEASONS

Floods can happen all year round, but they are more likely in some seasons than others. They happen in the seasons when most rain falls.

Rainy Seasons

In places with four seasons a year, such as the United States, floods happen from spring to fall, when there is more rain. Tropical countries lie close to the middle of the planet. The weather is warmest there and most water rises into the atmosphere. The clouds that bring wet seasons are called the tropical rain belt. Tropical countries, such as India, have two seasons. They are the **monsoon season** and the dry season. In the monsoon season, it can rain all day, making flooding more likely.

From June to September, parts of India become the wettest places in the world, making flooding common.

Where in the World?

The United States has four seasons but, between July and mid-September, there is a North American monsoon. During that time, there are more thunderstorms and rain over large parts of the southwestern United States and northwestern Mexico. Winds flow from moist ocean areas into dry land areas. During the day, the heat builds up and heavy rains usually occur during the late afternoon or early evening. This rainy weather increases the risk of flooding. After September, the winds change direction and blow from land to sea. The land cools and the weather becomes drier.

The North American monsoon brings heavy rainstorms that cause serious floods.

Bringing Rain

It rains very hard for months in the monsoon season. Winds carry moisture over the land from the warm oceans nearby. The strongest monsoons are in India, Bangladesh, Nepal, Pakistan, and Sri Lanka. Warm air rises over the hot land in summer and cool, wet air from the ocean blows in to fill the space left behind. Those winds bring heavy rain. The rains are vital for watering crops. After a long dry season, they bring around 80 percent of the country's total **annual** rainfall. Without the monsoon rains, crops would fail, and people could starve. Sometimes, the heavy rains can flood towns and villages and destroy crops. The rain gradually fills rivers, making them wider than normal, until they overflow.

COASTS UNDER ATTACK

At the coast, waves usually cause little damage—they roll gently onto the beach, or crash against the cliffs. However, coastal flooding is a problem when seawater suddenly sweeps inland. Then towns and villages on the coast can be flooded by storm surges and tsunamis.

Storm Surges

Waves form on the ocean's surface when wind blows over the surface of the water. The stronger the wind, the bigger the waves. Waves also get bigger as the wind blows them over a longer distance. A storm surge happens when strong winds push large waves toward the coast when there is a high **tide**. That combination can be disastrous. Storm surges can cause much flooding in areas of low-lying land. Large waves can pour over sea walls and smash roads, buildings, and railroads. Storm surges are mostly caused by intense storms called tropical cyclones. They are also known as typhoons or hurricanes.

Cyclone Debbie caused a dangerous storm surge and high-speed floodwaters in Queensland, Australia, in 2017.

WHIRLING STORMS

Tropical cyclones are storms with very strong winds. These winds spin in a circle and bring heavy rainfall. Cyclones form over warm, tropical oceans. The center of a cyclone is called the eye. The eye is surrounded by a ring of clouds called the eye wall. This is where the winds are strongest. Around the eye wall are the clouds that bring rain. Cyclones can cause storm surges at the coast. The heavy rainfall they bring can cause flooding.

This photograph was taken in 2004 in Phuket, Thailand, as the second wave of a tsunami struck the coast.

Tsunami Strikes

A tsunami is a series of waves that can hit coastlines as tall walls of water. Tsunamis keep moving very quickly across land. Most tsunamis happen when an earthquake shakes the ocean floor, which moves large amounts of water suddenly. That creates a series of waves at the ocean's surface. Tsunami waves can be 130 feet (40 m) higher than the normal level of the sea. They hit places along a coast with the same force as a wall of concrete. They can flood wide areas of land very fast. The water can carry boats, vehicles, and broken parts of buildings for miles inland. People, animals, buildings, and farms become covered in deep, dirty water.

Disaster Zone

Kerala, India

In the state of Kerala, southern India, about 90 percent of the rain falls during the monsoon months. In 2018, the rain was much heavier than usual and terrible flooding took place.

Mountain roads were heavily damaged by the monsoon and floods in Kerala in 2018.

Deadly Rainfall

Heavy rain started to fall in Kerala on August 8, 2018. It was the monsoon season, but these rains were almost twice as heavy as usual. The extra water soaked Kerala's steep and rolling hills, then it flowed quickly into the rivers. Several rivers spilled over their banks—one river's water flooded 40 villages. In some villages, floodwaters up to 10 feet (3 m) high entered homes. Helicopters airlifted families from the rooftops of their homes. Survivors, who were holding onto floating debris, were rescued from the dirty waters.

Landslide Disaster

The downpours triggered not only catastrophic floods, but also **landslides**. The landslides killed many people, and collapsed homes and bridges. Some experts say that the landslides were made worse by deforestation. Trees cut from the slopes left behind bare soil. The heavy rains soaked and loosened the soil. When soil is soaked, it becomes too heavy to rest on a slope and slides down onto land below. Some parts of Kerala may also have suffered worse floods because wetlands and lakes had been drained and built on. Their job was to soak up the monsoon rains.

Damage and Devastation

The floods covered a vast area of land and buildings. Whole housing zones were swept away, and large areas of farmland were flooded. Crops such as coffee, rubber, tea, and black pepper were ruined. Kerala lost more than $100 million in destroyed crops. About 1 million people were made homeless. More than 350 people died, either by drowning or as a result of mud and falling rocks from landslides.

Q People argue that we need to respect our **environment** to prevent floods. What do you think they mean by this?

DAM DAMAGE

Some experts say that the local authorities actually made the floods worse. The authorities emptied Kerala's dams during the monsoon when they were full of water. The experts argue it was more sensible to empty the dams slowly during the dry season instead of during the monsoon. Waiting for the dams to fill up was a mistake, because then too much water was let go at once, adding to the flooding from the monsoon.

Volunteers worked together to help rescue victims of the terrible floods.

Chapter 4

SCIENTISTS ON THE CASE

Scientists study floods to try to better understand them. It is important to figure out where and when a flood might happen because that knowledge will give people extra time to get away from the floodwaters.

Studying Earth from Space

Satellites are objects that are sent into **orbit** to circle our planet. They send images and information to computers on Earth. Satellites are used to study the stars. Some satellites carry Internet signals. Others keep track of the weather on Earth by watching it from above. Weather satellites have cameras that take images of Earth's atmosphere. Those images show scientists where thick clouds that might bring rain are forming. A transmitter onboard the satellite sends the images and other information to computers on Earth, where scientists can analyze it.

Night Sight

To study clouds during the night, satellites have to use special **sensors** that measure heat. Because there is no sunlight at night, the sensors measure the heat the clouds give off. **Infrared images** show where clouds are because they are always colder than the land and water below. Clouds with very cold tops in images are the most dangerous because they bring thunderstorms.

Watching Water Vapor

Satellite images can also be used to learn how much water there is in the upper parts of Earth's atmosphere. Water vapor images are useful to show where heavy rain is possible. The more water vapor there is in the air, the heavier the rains that might fall. Water vapor satellite pictures show wetter areas as white, while dry regions will be dark.

In satellite pictures taken during the day, clouds show up as white because they **reflect** the light from the sun. The ground is normally gray to green, and water is dark.

SATELLITES IN SPACE

Weather satellites in **geostationary** orbits stay in one place above Earth's surface. They sit 22,400 miles (36,050 km) above the **equator**. Other satellites circle over the North Pole to the South Pole and back again. They can travel as close to the surface as 500 miles (805 km). They take detailed pictures of weather patterns all over the world.

Satellites in space are used to build weather maps. These can help identify when and where there is a risk of flooding.

THE LAY OF THE LAND

Scientists carefully study the land to figure out where it might flood. A flood zone is the area of land that will flood when there is river or coastal flooding. The scientists need to collect a lot of information about what makes a flood zone.

Learning About the Land

Scientists collect detailed information about areas of land. They want to figure out how big and damaging a flood might become. They need to know how damp the soil is, what temperature the ground is, and if there is any snow on the ground. They also need to know if the land is flat. Is it the same level as the sea, or steep and sloping? They also need to figure out what types of plants cover the soil and how many plants there are. The type of rock is also important. Scientists study the rock to see if it is hard and solid, or if it has pores, or little holes, in it.

Scientists take and analyze soil **samples** to help them make maps. The maps are used to show where floods are likely to happen.

MAPPING FLOODS

In the United States, flooding happens in all 50 states. There are nearly 12.5 million square miles (32 million sq km) at risk, but how do people know which areas are in the most danger? Flood risk maps, such as this one, use colors to show the different levels of risk in flood zones. For example, on a flood map, blue is usually used to show the normal water surface. Red shows the areas where floodwaters usually reach. Yellow shows the area where flooding is less common, but can happen.

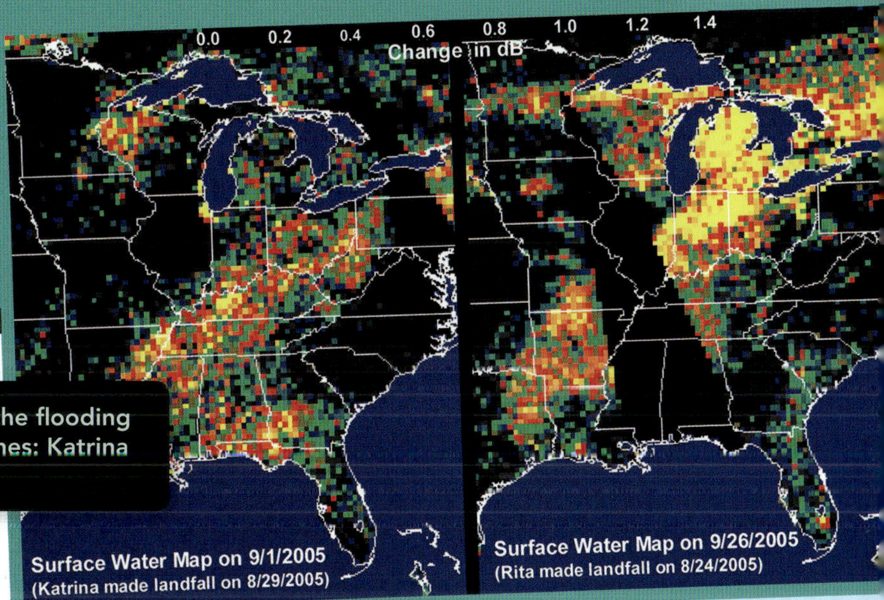

These maps compare the flooding caused by two hurricanes: Katrina (left) and Rita (right).

Making the Difference

Small differences in the land can make a big difference to its flood risk. For example, hard rock does not let water in, so it is at more risk of runoff and floods. Porous rock absorbs the rainwater. Steep hillsides that let rainwater rush quickly down their sides increase the risk of floods below. A small 6-inch (15 cm) deep creek in the mountains can grow to a 10-foot- (3 m) deep raging river in less than an hour during a heavy thunderstorm. The more information scientists have about the landscape, the more help maps of the areas will be. Those maps show which areas are most likely to be flooded when heavy rainfall hits.

Disaster Zone

North Carolina

On September 14, 2018, Hurricane Florence hit. It rained and rained and rained across parts of North and South Carolina, causing terrible floods.

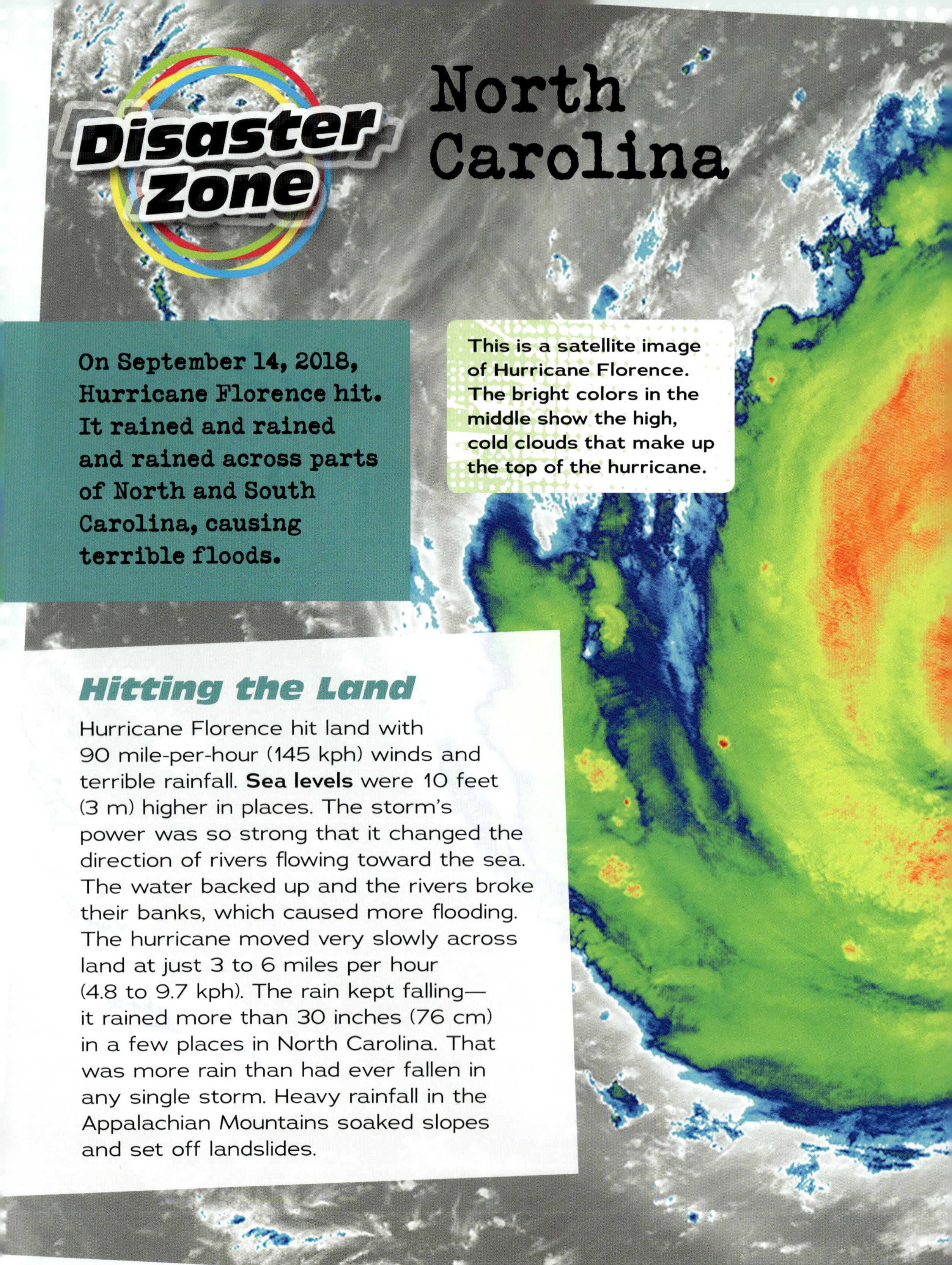

This is a satellite image of Hurricane Florence. The bright colors in the middle show the high, cold clouds that make up the top of the hurricane.

Hitting the Land

Hurricane Florence hit land with 90 mile-per-hour (145 kph) winds and terrible rainfall. **Sea levels** were 10 feet (3 m) higher in places. The storm's power was so strong that it changed the direction of rivers flowing toward the sea. The water backed up and the rivers broke their banks, which caused more flooding. The hurricane moved very slowly across land at just 3 to 6 miles per hour (4.8 to 9.7 kph). The rain kept falling—it rained more than 30 inches (76 cm) in a few places in North Carolina. That was more rain than had ever fallen in any single storm. Heavy rainfall in the Appalachian Mountains soaked slopes and set off landslides.

Disaster Damage

Hurricane Florence brought so much rain that flooding rather than strong hurricane winds caused most of the damage to buildings. The US National Weather Service (NWS) said the hurricane dropped around 8 trillion gallons (30 trillion l) of water. Record river flooding destroyed roads, and thousands of homes and businesses were damaged. Water also flowed off the slopes to make the floods below even worse. Nearly 1 million homes lost power. More than 5,200 people had to be rescued from the flood. The cost of that extreme weather event was put at about $22 billion. Altogether, 51 people died because of the hurricane.

Flooding caused by the Hurricane Florence storm surge covered streets and buildings in North Carolina.

SCIENCE PREDICTION

Satellite images of the atmosphere over the Atlantic Ocean showed heavy storm clouds as early as September 1. Within five days, the storm had grown into a powerful hurricane named Florence. By studying its strength and the speed and direction of its winds, Florence's threat was spotted early. More than 1 million people left their homes on the coast before of the hurricane reached land at Wrightsville Beach in North Carolina on September 14.

Chapter 5

SUPER FLOODS OF THE FUTURE

Super floods have happened often in Earth's history. However, there is a risk there may be more super floods in the future. Some scientists fear that as Earth's climate changes, all floods may become bigger, more frequent, and more deadly.

Climate Change

When people talk about climate change, they mean a general rise in temperature. The world's temperatures have risen more quickly in the past 140 years than in the past 1,400 years. Scientists believe that climate change is happening today because of how we live. When people burn **fossil fuels** such as coal, oil, and natural gas, they add greenhouse gases, such as carbon dioxide, to the air. Those gases trap heat in Earth's atmosphere, as they do in a glass greenhouse, making the planet warmer.

More Rain

Hotter temperatures mean more water evaporates into the air. The warmer atmosphere is able to hold more water vapor. Scientists have shown that there is more evaporation from oceans as the world gets warmer. That means more rain will bring more floods. As the seas warm up, there is also more chance of powerful hurricanes happening more often. Hurricanes begin over warm oceans. They bring heavy rains and can cause storm surges, too.

Higher Sea Levels

As global temperatures increase, ice sheets at the North and South Poles are melting into the sea. This makes sea levels rise. In the past 100 years, sea levels have risen by 8 inches (20 cm) and are still rising. When sea levels are higher, they make storm surges, big waves, and high tides more likely. They flow over beaches and the tops of sea walls to flood dry land at the coasts.

BLAMING CLIMATE CHANGE

It is impossible to blame climate change for every flood. Floods happen for many different reasons,and there have always been extreme weather events. What we can say for sure is that changes in weather do lead to an increased number of natural disasters.

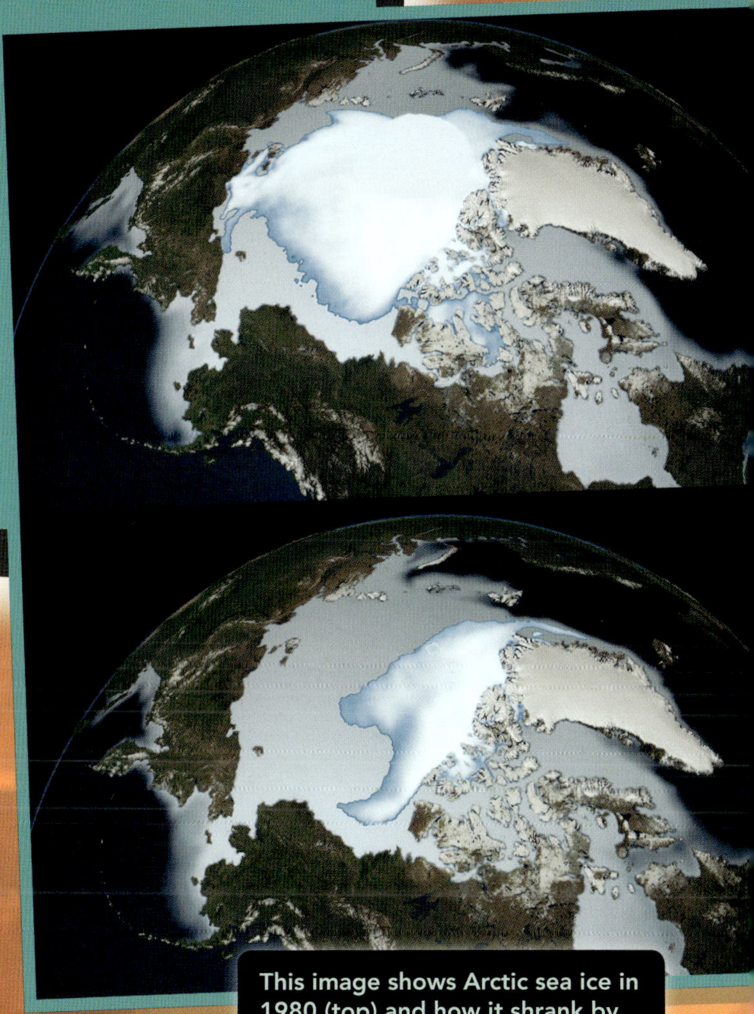

This image shows Arctic sea ice in 1980 (top) and how it shrank by 2012 (bottom) as a result of the global rise in temperatures.

Could our use of fossil fuels, which causes climate change, increase the risk of more and stronger floods in the future?

MAPPING SUPER FLOODS

Scientists make models and maps to try to understand super floods. The models show how and where there may be a flood in the future. Super floods are rare, however. That can make it difficult to learn about them, even though they cause so much damage.

Secrets in the Sediment

There is information in the layers of debris that super floods leave behind. Floodwaters leave layers of sand, mud, and other tiny bits of material as they drain away. This is called sediment. Over hundreds or thousands of years, sediment mixes with other materials. Separating the materials can be difficult. When scientists studied ancient super floods from ice-dammed lakes, such as Glacial Lake Missoula, they saw that sediment left from this type of flood contained certain materials. These helped them figure out just how far the floodwaters had traveled.

Studying sediments can reveal secrets about super floods in the far-distant past.

Mapping the Land

The shape of the land also tells us about past super floods. Scientists have studied Great Britain. During the last ice age, more than 200,000 years ago, a super flood cut it off from **continental** Europe. Scientists have made a map of the seafloor around Great Britain from information collected by special ships with **sonar devices**. The images showed how the floodwaters had changed the seafloor. The flood had made a straight, wide underwater valley when the water and moving rocks dug out deep channels. Areas of raised seafloor separated deeper channels. They looked like shapes created 15,000 years ago by the Glacial Lake Missoula flood.

FLOODS ON MARS

Studies of the planet Mars have proven that there were also super floods on planets other than Earth. **Radar** and other instruments exploring the surface of Mars are studying channels carved by an ancient mega flood on the planet. Today, there is no liquid water—and no floods—on Mars.

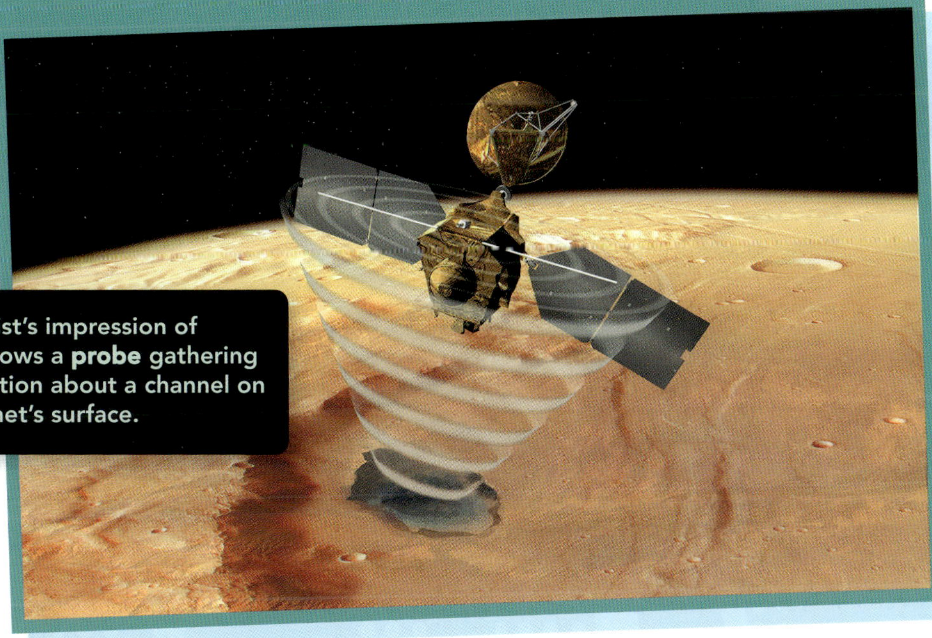

This artist's impression of Mars shows a **probe** gathering information about a channel on the planet's surface.

Disaster Zone

California

Experts learn from studying past super floods. They learn how often and why they happened. Some scientists believe there could be a super flood in California in the future and that this super flood would destroy the state.

In 2018, flooding led to huge mudslides in Santa Barbara. They damaged buildings and destroyed vehicles. However, this would be just a fraction of the damage caused by a future super flood.

Fears for the Future

Scientists have found **evidence** about super floods in California. Heavy rainfall hits California every 100 to 200 years, caused by narrow bands of water vapor in the atmosphere. Those bands sit about 1 mile (1.6 km) above the ocean and are thousands of miles long. They can cause storms that drop more rain than we could ever imagine, resulting in floods. It is likely that such a flood will happen in California, and it could be within the next 40 years. Almost certainly, it will happen by the end of the century.

Mega Disaster

If another mega storm happens, enormous amounts of rain will fall in a very short time. The water will flow down slopes and across the landscape. It will overflow streams and rivers. California's Central Valley will turn into a huge inland sea. The storm could bring hurricane-force winds and cause hundreds of landslides. These would destroy roads, highways, and homes. Around 1.5 million people would have to leave their homes and move to safety. Floodwaters would cover about one-quarter of the state, and much of the region from Silicon Valley to Los Angeles would be destroyed.

FUTURE DISASTER

A future super flood could be like those of 1861 and 1862 when terrible floods followed very heavy storms. Those storms dropped about 10 feet (3 m) of rain in parts of California. Central and southern California were under water for up to six months, and the water reached depths up to 30 feet (9 m). Today, those same areas are home to California's fastest-growing cities, and more people are now at risk if a super flood happens.

California's next super flood could be worse than eight Hurricane Katrinas. It could become the world's first trillion-dollar natural disaster.

Q How might you persuade people to prepare for future mega floods? What arguments would you use?

Chapter 6

BE PREPARED

Floods are a force of nature. They cannot be stopped. We cannot stop rain from falling. We cannot stop runoff from causing riverbanks to burst. They are natural events. However, we can do something to stop them from having great impact.

Water Barriers

In some coastal regions, sea walls are built. The walls keep storm surges from pushing seawater onto land. Inland, walls and levees are built to hold back water. Lakes, dams, and reservoirs are constructed to hold extra water when a flood happens. Detention basins are small, human-made lakes. They connect to waterways such as rivers. They can collect and temporarily store floodwaters. If a flood happens, water flows into drains, then to the detention basin. That helps stop flooding downstream, and gives people nearby more time to get away.

Safer Cities

In some places, people are moving their homes away from flood zones. They are building new homes in safer places. As sea levels rise, people on some **flood-prone** shores are moving farther inland. People are also planting trees, shrubs, and grasses in flood zones beyond cities. Those will help protect the land from **erosion**. The plants help soak up excess water in case of a flood. People can also make cities safer by not dropping garbage. Garbage blocks drains and stops water from flowing quickly through them, which increases the chance of flooding.

HOMES THAT FLOAT

To make their homes floodproof, some people are building floating houses. They sit on **pontoons** that float. The houses are built on land and will only float when there is a flood. If a flood happens, the homes will rise with the water and not be covered by it.

The Thames Barrier in London, England, is designed to keep central London from being flooded by storm surges.

Better Buildings

If people build near a river or an ocean, it helps if the building is higher than the height of floodwaters. Some people build houses on stilts. Others build on raised platforms on banks of earth or concrete. Large office buildings sometimes put a parking lot underneath. The walls have openings to let floodwaters flow straight through the ground floor and back outside.

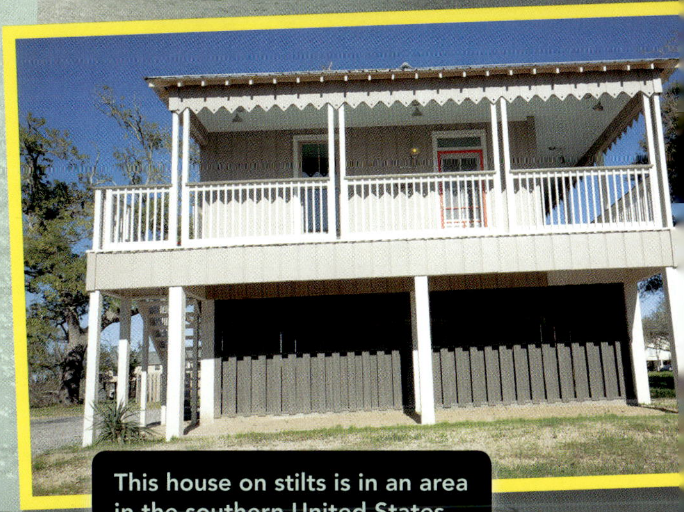

This house on stilts is in an area in the southern United States that is prone to flooding.

FLOOD-WARNING SYSTEMS

It is important to be prepared. We need to know what flood warnings are, and what to do when there is a flood warning.

Flood Watch or Warning?

Flood watches and flood warnings are official announcements. Authorities make them on television and radio, or by SMS, phone, and e-mail. A flood watch means flooding is possible. During a flood watch, people must check for updates. They must be ready in case they have to leave their homes.

A flood warning means there is flooding or there soon will be. It warns people to move to higher ground, and that they must act immediately to avoid drowning and lessen the damage to their homes.

Action to Take

When a flood watch is given, everyone must remember the evacuation plans and instructions. They must know where their local shelters are. They need to know what routes to take to get to higher ground and safety. Families should make sure they know emergency numbers and important information, and should also check their emergency kits. An emergency kit is a waterproof storage container with items people might need if they leave home in a hurry. Emergency kits can save lives.

The best way to be safe in a flood is for people to recognize warnings and obey them immediately.

Following Warnings

People must act as soon as a flood warning is given because flash floods happen in a very short period of time. People should make sure all their valuable things are safe, such as pets and family photo albums. All electrical appliances, gas, and heating supplies should be turned off. Items should be taken upstairs or to a safe place, if there is time. People should then leave quickly, while keeping safe.

It is best to leave an area before a flood hits, rather than stay and have to escape after.

PLANNING FOR AN EMERGENCY

An emergency kit should have things such as a battery-operated radio and a flashlight, both with spare batteries. A first aid kit, candles, and waterproof matches should be included. Important papers, including emergency contact numbers, and any medicine people need should be packed. Enough food and bottled water for three days should be also included, along with warm clothing and blankets.

SURVIVING A SUPER FLOOD

The thought of dangerous super floods can be scary. While it is important to take floods seriously, there are also things people can do to manage the risk. Being prepared can help people survive a flood.

Keep Safe

In a flood, the main thing to remember is that the safest place to be is on higher ground away from the water. People should get to higher ground as quickly as they can. If possible, they should avoid floodwater because it may contain dangerous chemicals and disease-causing germs. Also, moving water just 6 inches (15 cm) deep can knock people off their feet. If they have to walk in water, they should try to walk where the water is not moving. A stick can be used to check if the ground is solid ahead, and people should stay away from power lines.

If people are caught by a flood before they can escape, they should move upstairs with their family and pets. Any high place with a means of escape is good. Some people sit on the roof if they have to. There, they can stay above the floodwaters and where rescuers can see them.

Helicopter rescue teams are trained to rescue flood victims stranded on their roofs.

Hope for the Future

If people in flood zones do not take chances, they should be safe. They must listen to and obey evacuation warnings when they are given. Scientists are working hard to find ways of **predicting** where floods will happen sooner so they can give people plenty of time to get to safety out of the path of floodwaters.

Cleaning up after a flood takes time. However, if everyone works together, people can soon rebuild their lives and their homes.

AFTER A FLOOD

When a flood is over, rescue services spring into action. Everyone must listen for information and instructions from the authorities. No one should try to return to their home until they are told it is safe to do so. When returning home, they should be careful because there might be hidden dangers in the floodwater, such as sharp objects or snakes.

Glossary

annual yearly
atmosphere the blanket of gases surrounding Earth
bacteria tiny living things that can cause disease
canyons huge, deep cuts in the surface of Earth, often with rivers flowing through them
chaos a state of utter confusion or disorder
continental related to the continents, including Europe, North America, and South America
crops plants grown for food
debris pieces from something that has been destroyed
downstream farther along a river, in the direction in which it is flowing
drainage system a system for water to flow away into the ground or down pipes
drought a long period without rain
dysentery an infectious disease of the digestive system causing severe diarrhea
energy power translated into motion
environment the natural world around us
equator an imaginary horizontal line around the center of Earth
erosion the wearing away of soil or rock by wind, water, or ice
evacuated moved away from danger in an organized way
evidence proof of something
flood-prone likely to have flooding
fossil fuels oil, coal, and gas, formed from long-dead plants and animals
geostationary matching Earth's rotation at a fixed point in orbit
glaciers slow-moving rivers of ice
ice age a period when Earth was much colder and partly covered in thick ice
ice sheets large, thick areas of ice
infrared images a form of photography that shows the heat given off by objects
landslides large areas of mud, soil, and rock that slide down a slope
monsoon season the time of year at which monsoons occur
orbit the curved path through which objects in space move around a planet or star
plantation a very large farm
pontoons floats or flat-bottomed boats
power plants places in which power is created, often by burning fossil fuels
predicting telling what might happen in the future
pressure a pushing force
probe an uncrewed space vehicle used to learn more about space
radar a system that uses radio waves to locate and measure objects at a distance
reflect to bounce back
reservoir a human-made lake used to store water
samples small pieces of something
sea levels the average heights of the ocean's surface
sensors devices that are used to record properties such as heat, sound, and light
sonar devices machines that use sound waves to detect and measure objects underwater
tide the daily rise and fall of the ocean
viruses tiny organisms that are invisible to the naked human eye but can cause disease
volunteers people who offer to work on something for no pay
water vapor water in gas form
wetlands lands almost permanently covered with shallow water

Find Out More

Books

Brundle, Joanna. *Tsunamis and Floods* (Transforming Earth's Geography). KidHaven Publishing Company, 2019.

Hofer, Charles. *Washed Away by Floods* (Natural Disasters: How People Survive). PowerKids Press, 2018.

Kurtz, Kevin. *Climate Change and Rising Sea Levels* (Searchlight Books: Climate Change). Lerner Publishing, 2019.

Machajewski, Sarah. *Storms, Floods, and Erosion* (Spotlight on Weather and Natural Disasters). PowerKids Press, 2019.

Websites

Read up about floods at:
www.nationalgeographic.com/environment/natural-disasters/floods

Find out more about floods and flood risks at:
www.nssl.noaa.gov/education/svrwx101/floods

Learn more about how to prepare for floods at:
www.ready.gov/floods

Publisher's note to educators and parents:
All the websites featured above have been carefully reviewed to ensure that they are suitable for students. However, many websites change often, and we cannot guarantee that a site's future contents will continue to meet our high standards of educational value. Please be advised that students should be closely monitored whenever they access the Internet.

Index

About the Author

Louise Spilsbury is an award-winning author who has written hundreds of wonderful books for children. She has written many books about natural disasters and has a healthy respect for these awesome forces of nature!